TINY CACTUS PUBLISHING

CLORIST NAME

TEST PAGE
PREPARE YOUR COLOR

WARM UP !!

BITCH

FUCKWIT

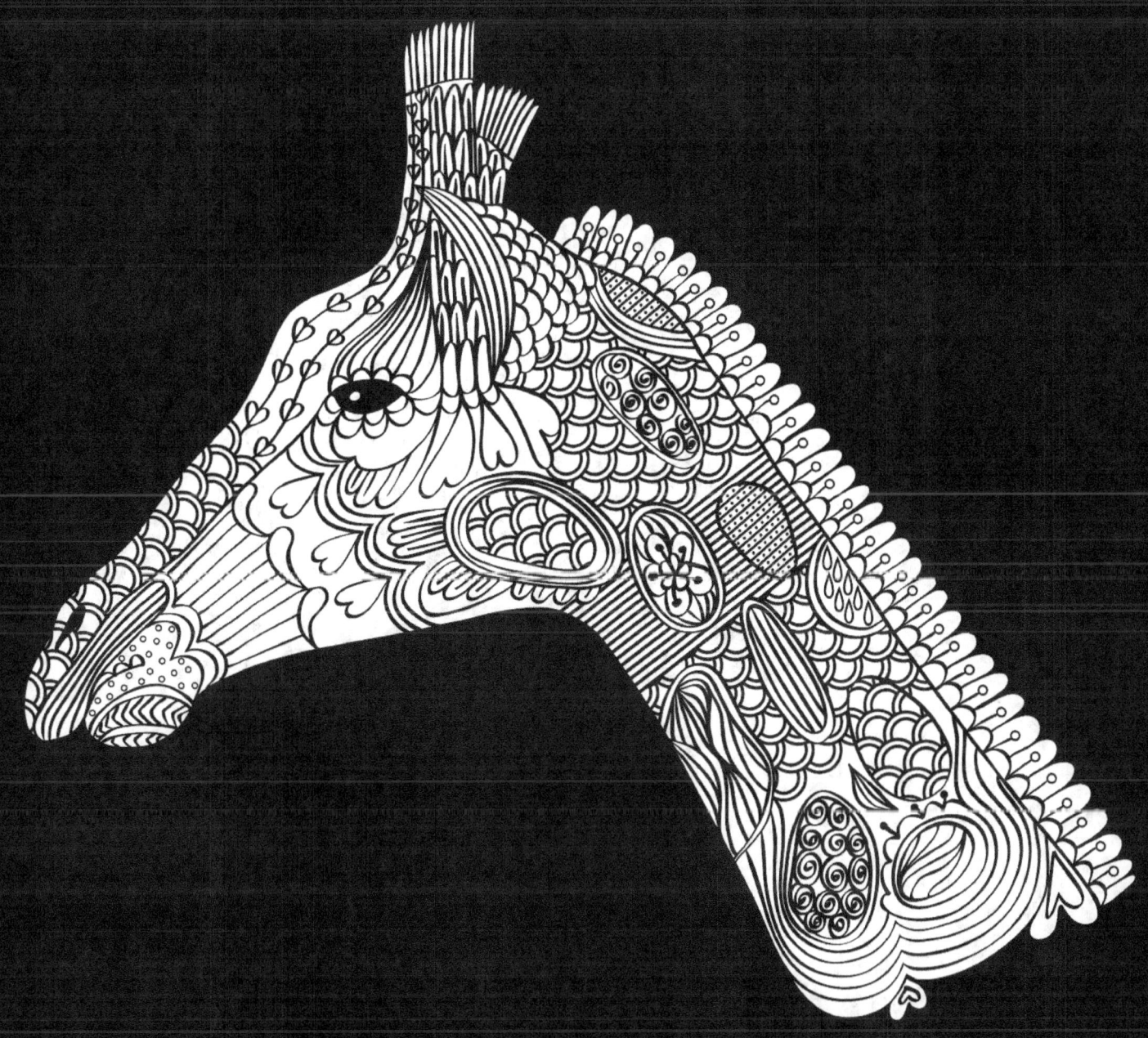

BITCH FACE

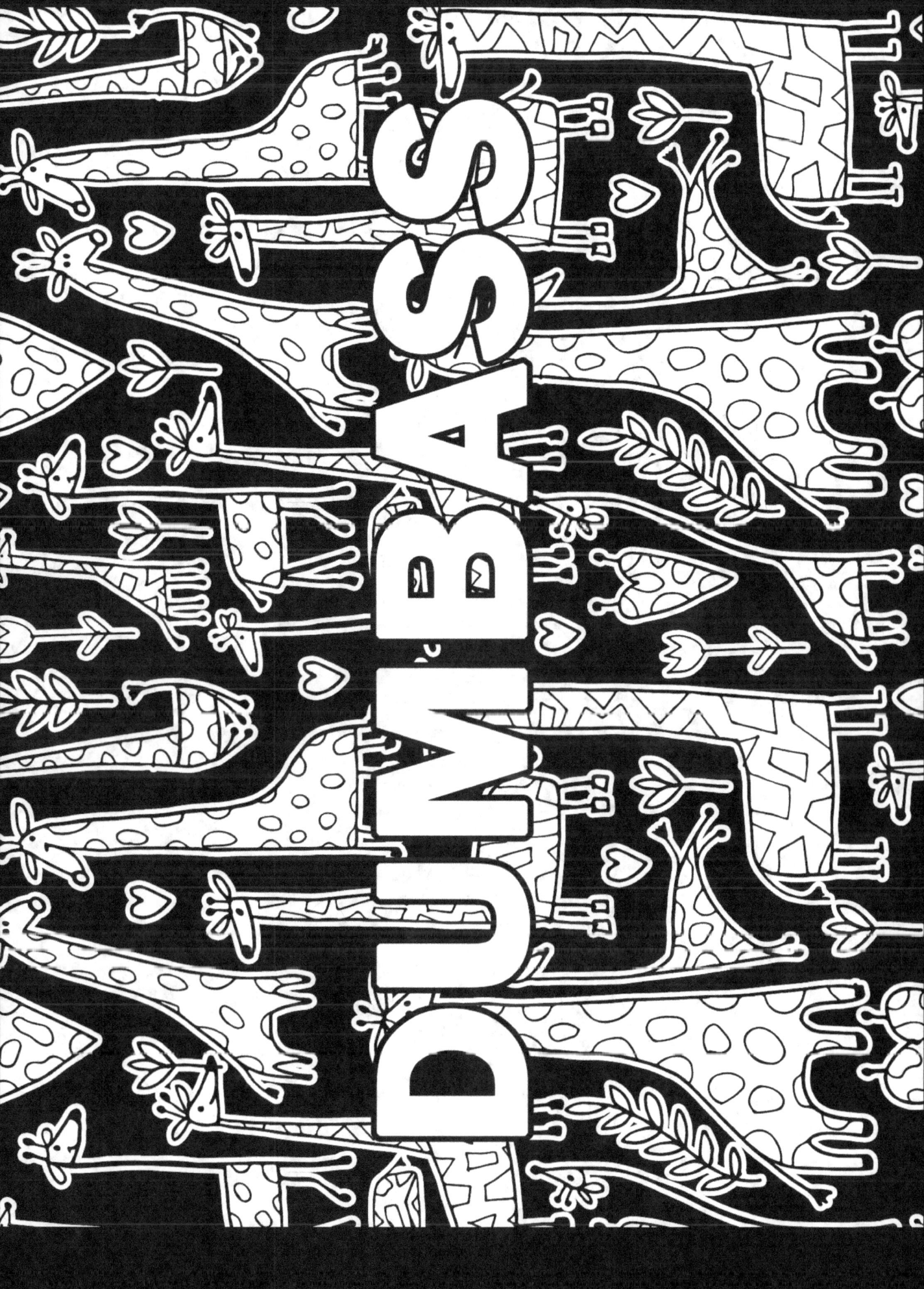

ASSHAT

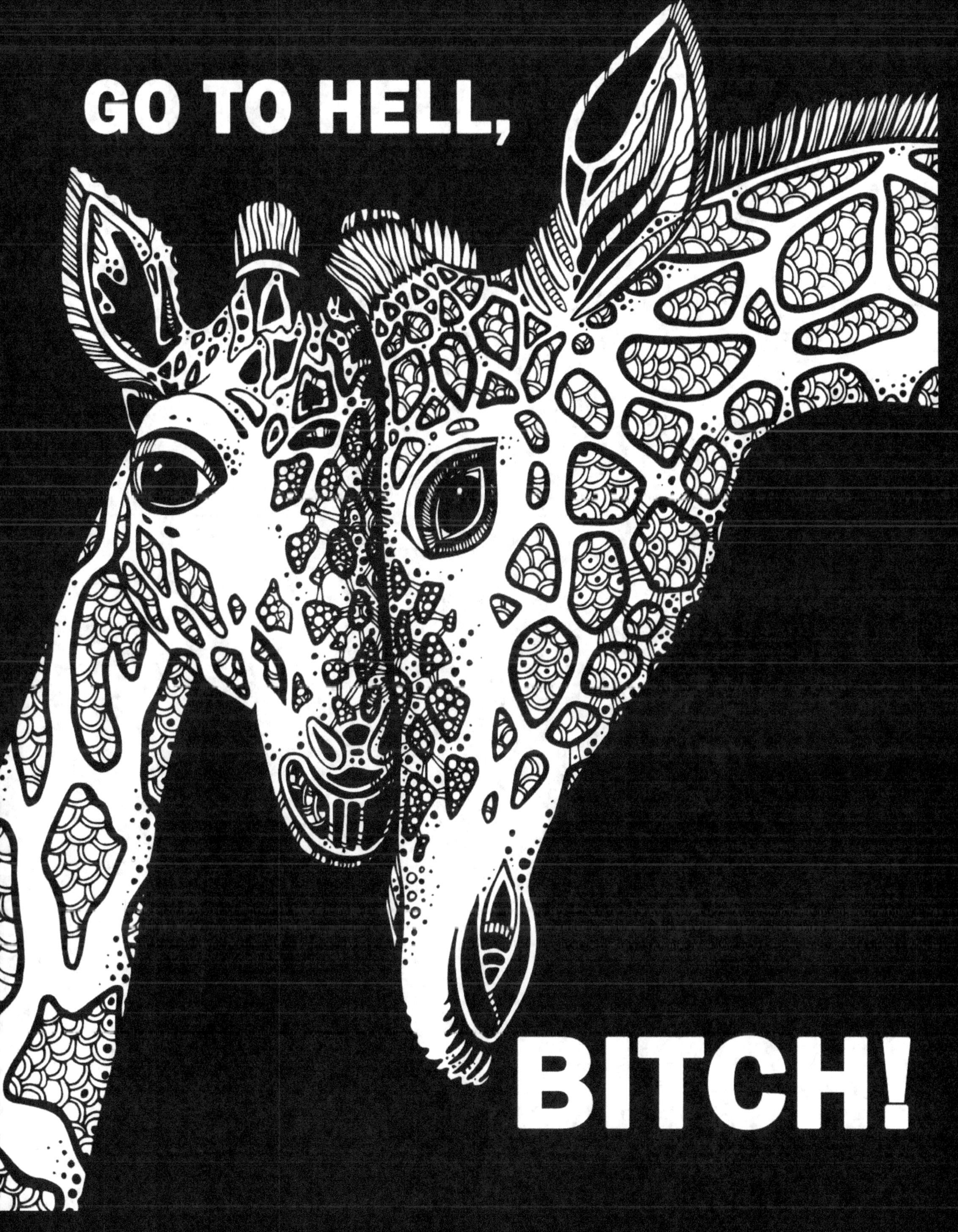
GO TO HELL,
BITCH!

BULLSHIT

CAMN

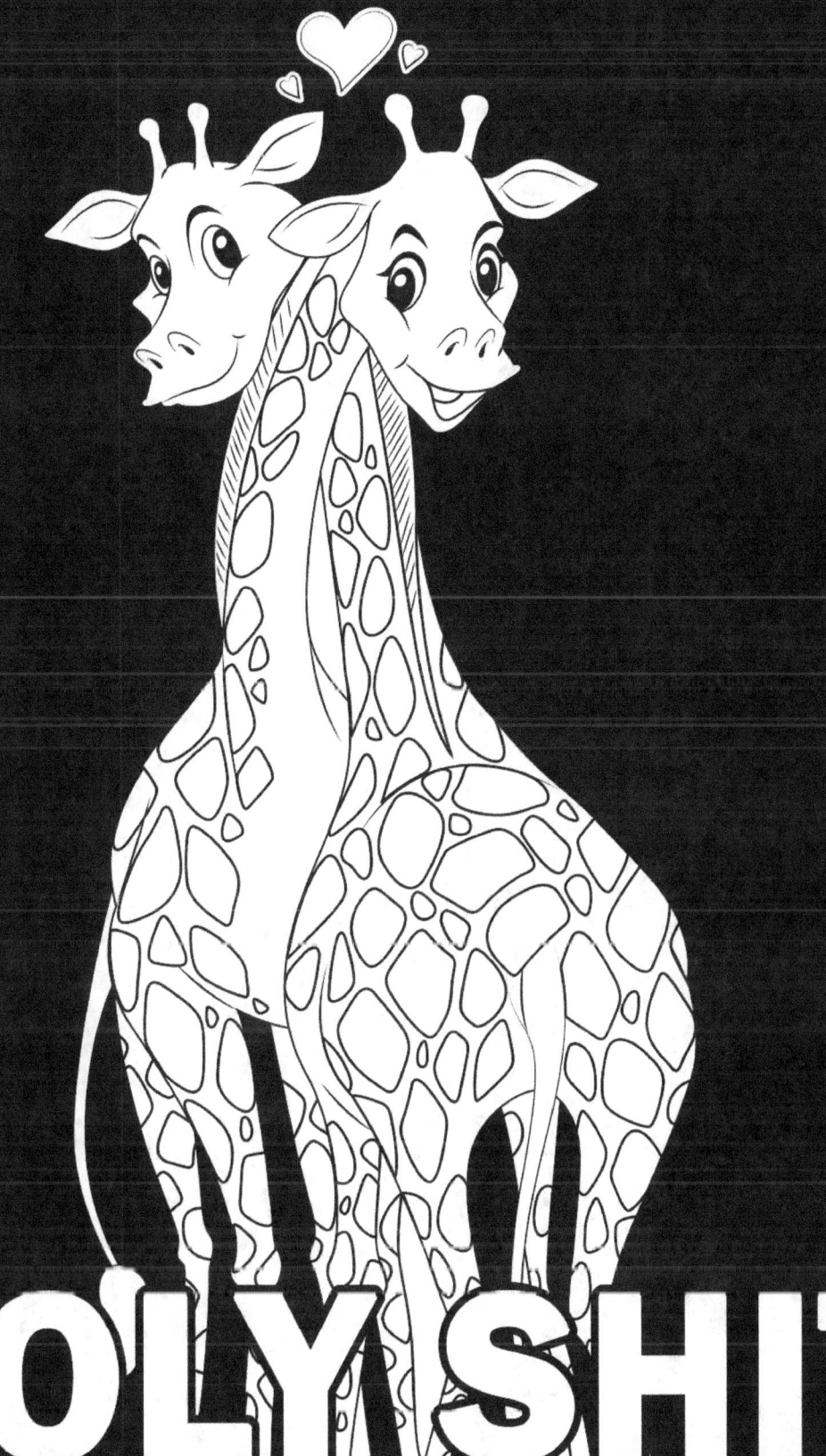

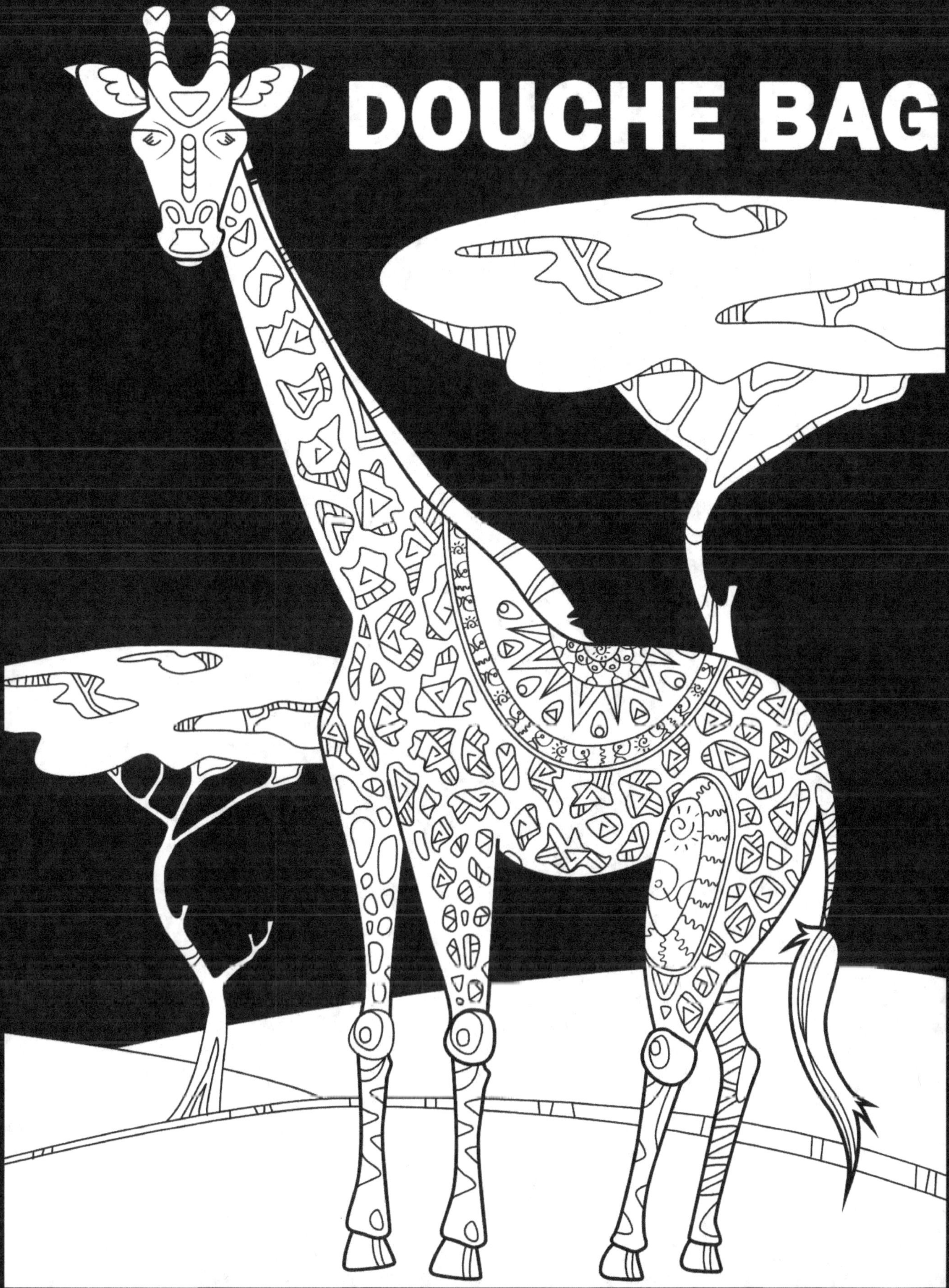

DICKHEAD

FUCK THAT SHIT

ASSFACE

BASTARD

FUCK YOU

YOU'RE SUCH A DICK

DUMB
PIECE
OF
SHIT

HOLY SHIT

www.ingramcontent.com/pod-product-compliance
Lightning Source LLC
Chambersburg PA
CBHW080132240526
45468CB00009BA/2381

978 1 985 054 4 1 7